Math Is Everyw[here]

MATH AT [THE]

PARK

By Claire Romaine

Gareth Stevens
PUBLISHING

Please visit our website, www.garethstevens.com. For a free color catalog of all our
high-quality books, call toll free 1-800-542-2595 or fax 1-877-542-2596.

Library of Congress Cataloging-in-Publication Data

Names: Romaine, Claire, author.
Title: Math at the park / Claire Romaine.
Description: New York : Gareth Stevens Publishing, [2017] | Series: Math is
  everywhere!
Identifiers: LCCN 2015045884 | ISBN 9781482446234 (pbk.) | ISBN 9781482446173 (library bound) | ISBN
9781482445961 (6 pack)
Subjects: LCSH: Mathematics–Juvenile literature.
Classification: LCC QA141.3 .R66 2017 | DDC 510–dc23
LC record available at http://lccn.loc.gov/2015045884

First Edition

Published in 2017 by
Gareth Stevens Publishing
111 East 14th Street, Suite 349
New York, NY 10003

Editor: Therese Shea
Designer: Sarah Liddell

Photo credits: Cover, p. 1 Alpha Zynism/Shutterstock.com; p. 5 T photography/Shutterstock.com; pp. 7, 24
(fountain) Pavel L Photo and Video/Shutterstock.com; p. 9 Air Images/Shutterstock.com; pp. 11, 24 (bench)
Ozgur Coskun/Shutterstock.com; p. 13 StanislavSukhin/Shutterstock.com; pp. 15, 24 (sunflower) Mike Vande
Ven Jr/Shutterstock.com; p. 17 Bocos Benedict/Shutterstock.com; p. 19 Lev Savitskiy/Shutterstock.com;
p. 21 Nadezhda1906/Shutterstock.com; p. 23 michaeljung/Shutterstock.com.

Printed in the United States of America

CPSIA compliance information: Batch #CS16GS: For further information contact Gareth Stevens, New York, New York at 1-800-542-2595.

Contents

We walk to the park.
The park is a
big rectangle!

We visit the fountain.
It looks like a circle.

We play hide-and-seek.
I count to five. One, two,
three, four, five!

My friend Annie hides
under a bench.

My friend Emma hides
behind a tree.

We visit the garden.
I find the tallest sunflower.

I see two dogs.
The white dog is bigger.

We visit
the playground.
We see numbers
in squares.

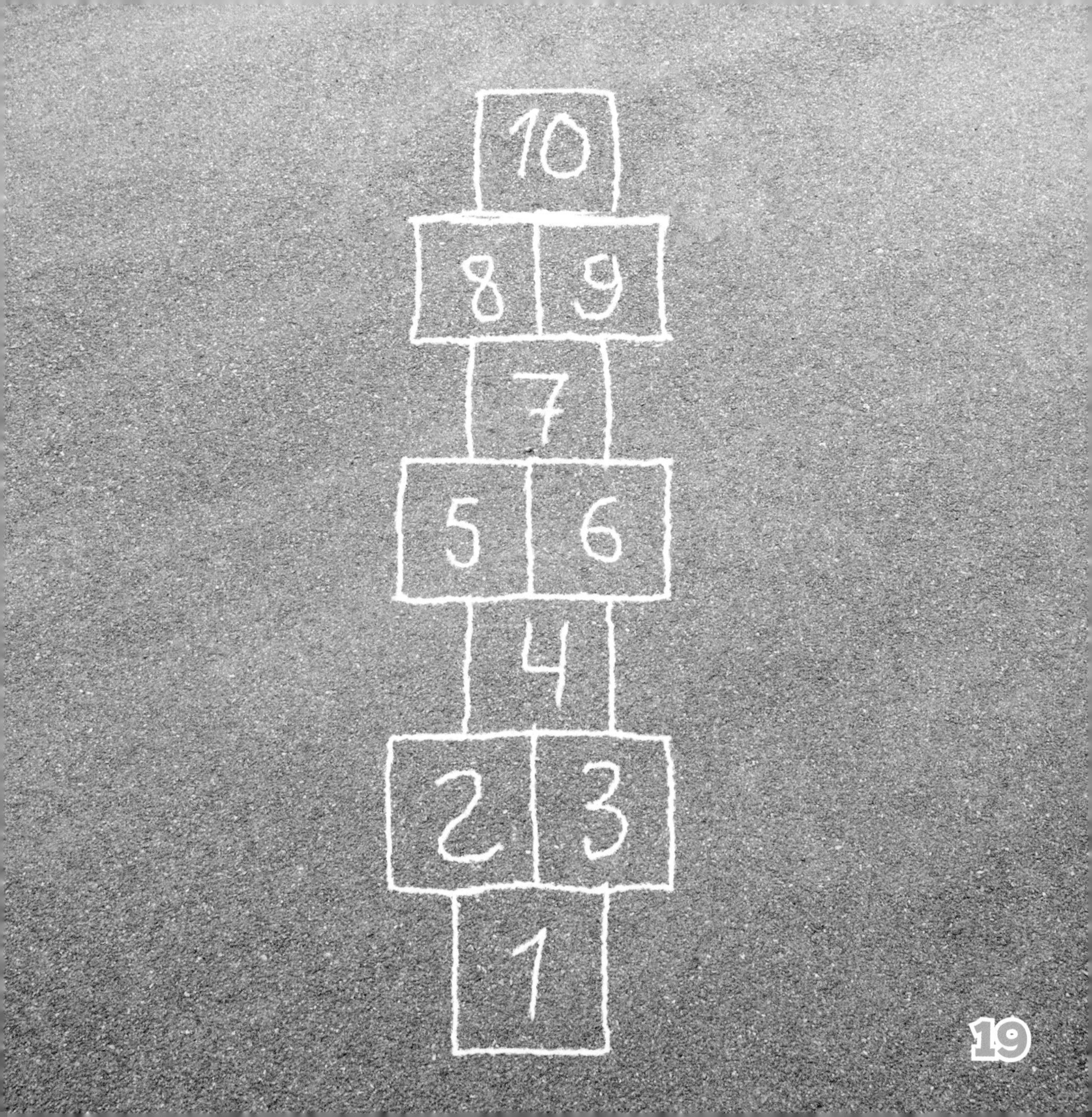
10
8
9
7
5
6
4
2
3
1

Let's play hopscotch!
I hop on the squares.

It is time to leave.
We had fun
at the park!

Words to Know

bench

fountain

sunflower

Index